Climate Pioneers: Leading the Charge for a Greener Future

Innovations, Policies, and Public Action in the Climate Crisis

Adriana J. Kidd

All rights reserved. No part of this publication may be reproduced, distributed, or transmitted in any form or by any means, including photocopying, recording, or other electronic or mechanical methods, without the prior written permission of the publisher, except in the case of brief quotations embodied in critical reviews and certain other noncommercial uses permitted by copyright law.

Copyright ©Adriana J. Kidd, 2024

TABLE OF CONTENT

CHAPTER ONE: INTRODUCTION TO CLIMATE CHANGE

    1.1 The Science of Climate Change

    1.2 Historical Context and Evolution of Climate Awareness

    1.3 Global Impact and Urgency

CHAPTER TWO: THE ROLE OF MEDIA IN CLIMATE

    2.1 Traditional Media Coverage

    2.2 The Rise of Digital Media

    2.3 Social Media's Influence on Public Perception

    2.4 Case Studies of Media Campaigns

CHAPTER THREE: EDUCATION AN CLIMATE LITERACY

    3.1 Integrating Climate Change into School Curriculums

    3.2 Higher Education and Research Initiatives

3.3 Community Education Programs

3.4 Online Courses and Resources

CHAPTER FOUR: GRASSROOTS MOVEMENTS AND ACTIVISM

4.1 The Role of Non-Governmental Organizations (NGOs)

4.2 Youth-Led Movements (e.g., Fridays for Future)

4.3 Local and Global Climate Marches

4.4 Case Studies of Successful Activism

CHAPTER FIVE: GOCERNMENT AND POLICY INITIATIVES

5.1 National Climate Policies

5.2 International Agreements (e.g., Paris Agreement)

CHAPTER SIX: CORPORATE RESPONSIBILITY AND CLIMATE ACTION

6.1 The Role of Corporations in Climate Change

6.2 Corporate Sustainability Initiatives

    6.3 Innovations in Clean Technology

    6.4 Case Studies of Corporate Leadership

CHAPTER SEVEN: INDIVIDUAL ACTIONS AND LIFESTYLE CHANGES

    7.1 Reducing Carbon Footprints

    7.2 Sustainable Consumption

    7.3 Renewable Energy and Home Improvements

    7.4 Community Involvement and Advocacy

    7.5 Education and Raising Awareness

CHAPTER EIGHT: THE FUTURE OF CLIMATE ACTION

    8.1 Emerging Technologies

    8.2 Global Collaboration and Policy Development

    8.3: Adapting to Climate Impacts

    8.4 Mobilizing Finance for Climate Action

8.5 The Role of Individuals in Shaping the Future

CHAPTER NINE: INNOVATION AND TECHNOLOGY IN CLIMATE SOLUTIONS

9.1 Renewable Energy Advances

9.2 Energy Storage Solutions

9.3 Smart Grid Technologies

9.4 Carbon Capture and Storage (CCS)

9.5 Innovations in Agriculture and Land Use

9.6 Case Studies of Technological Innovations

CHAPTER TEN: CLIMATE CHANGE AND PUBLIC HEALTH

10.1 Health Impacts of Climate Change

10.2 Mental Health and Climate Anxiety

10.3 Strategies for Mitigating Health Risks

10.4 Case Studies of Public Health Responses

CLIMATE ELEVEN: THE ECONOMICS OF CLIMATE ACTION

11.1 The Cost of Inaction

11.2 Economic Benefits of Climate Action

11.3 Market-Based Solutions

CHAPTER ONE: INTRODUCTION TO CLIMATE CHANGE

1.1 The Science of Climate Change

Climate change is defined as long-term shifts in temperature and weather patterns, primarily due to human activities. These shifts are largely driven by the increased concentration of greenhouse gases (GHGs) such as carbon dioxide (CO_2), methane (CH_4), and nitrous oxide (N_2O) in the atmosphere. The primary sources of these emissions include the burning of fossil fuels for energy, deforestation, and various industrial processes.

The science behind climate change involves understanding the Earth's energy balance. The sun's energy reaches the Earth, where some of it is absorbed and some is reflected back into space. GHGs trap some of this reflected energy, causing a warming effect known as the greenhouse effect. This warming leads to various changes in the climate system, including rising temperatures, altered precipitation patterns, and an increased frequency of extreme weather events.

Recent studies, such as those by the Intergovernmental Panel on Climate Change (IPCC), have provided comprehensive assessments of climate science, highlighting the unequivocal evidence of human influence on the climate. These studies show that without significant reductions in GHG emissions, global temperatures are expected to rise, leading to severe and irreversible impacts on ecosystems and human societies.

1.2 Historical Context and Evolution of Climate Awareness

The awareness of climate change has evolved significantly over the past century. Early warnings came from scientists like Svante Arrhenius in the late 19th century, who first quantified the warming effect of CO_2. In the mid-20th century, scientists began to observe increasing CO_2 levels and their potential impact on global temperatures.

The 1970s marked a turning point with the first Earth Day in 1970, which brought environmental issues, including climate change, to the forefront of public consciousness. The establishment of the United Nations Framework Convention on Climate Change (UNFCCC) in 1992 and the subsequent Kyoto Protocol in 1997 further highlighted the need for global cooperation to address climate change.

In recent years, the Paris Agreement of 2015 has been a significant milestone, with countries committing to limit global warming to well below 2°C above pre-industrial levels, and preferably to 1.5°C. This agreement has increased global awareness and action towards mitigating climate change.

1.3 Global Impact and Urgency

The impacts of climate change are far-reaching and profound. Rising temperatures have led to more frequent and intense heatwaves, affecting human health and agriculture. Changes in precipitation patterns have resulted in droughts in some regions and flooding in others, disrupting water supplies and damaging infrastructure.

Sea level rise, caused by the melting of polar ice caps and the thermal expansion of seawater, threatens coastal communities and ecosystems. According to the IPCC, global mean sea level has risen by about 20 cm since the start of the 20th century, with an accelerating rate in recent decades.

Biodiversity is also at risk as changing temperatures and habitats force species to migrate, adapt, or face extinction. Coral reefs, for example, are experiencing widespread bleaching events due to increased sea temperatures, threatening marine biodiversity and the livelihoods of communities that depend on them.

The urgency to address climate change is underscored by the potential for tipping points—thresholds beyond which certain impacts become irreversible. Examples include the collapse of the Greenland and West Antarctic ice sheets, which would result in substantial sea level rise, and the dieback of the Amazon rainforest, which would release large amounts of CO_2.

Overall, the need for immediate and sustained action to mitigate and adapt to climate change is critical to ensuring the well-being of current and future generations.

CHAPTER TWO: THE ROLE OF MEDIA IN CLIMATE

2.1 Traditional Media Coverage

Traditional media, including newspapers, television, and radio, has played a crucial role in disseminating information about climate change. Historically, coverage has varied in quality and quantity, often influenced by political and economic factors. For instance, significant climate events such as the release of IPCC reports or major international climate conferences often spur increased media attention.

High-profile publications like The New York Times, The Guardian, and National Geographic have consistently provided in-depth coverage of climate science, policy developments, and the impacts of climate change. Investigative journalism has also played a pivotal role in uncovering stories about the fossil fuel industry's role in climate change and efforts to downplay its significance.

2.2 The Rise of Digital Media

The advent of digital media has revolutionized how climate information is shared. Online news platforms, blogs, and video-sharing sites allow for more immediate and widespread dissemination of information. Digital media enables interactive and multimedia storytelling, making complex scientific data more accessible and engaging for the public.

Social media platforms, such as Twitter, Facebook, and Instagram, have become powerful tools for raising awareness and mobilizing action. Hashtags like #ClimateChange, #ActOnClimate, and #FridaysForFuture help amplify messages and connect people globally. Influencers and activists use these platforms to share their stories, promote campaigns, and engage with their followers.

2.3 Social Media's Influence on Public Perception

Social media's impact on public perception of climate change is significant. It allows for the rapid spread of information and can help build communities of interest and action. Viral campaigns, such as Greta Thunberg's school strike for climate, have gained international attention and inspired millions to take action.

However, social media also presents challenges, such as the spread of misinformation and the echo chamber effect, where users are exposed primarily to information that reinforces their existing beliefs. Efforts to combat misinformation, such as fact-checking initiatives and promoting credible sources, are essential for maintaining informed public discourse.

2.4 Case Studies of Media Campaigns

Successful media campaigns have demonstrated the power of coordinated efforts to raise awareness and drive action. National Geographic's "Planet or Plastic?" campaign, for example, has significantly raised awareness about plastic pollution and its impact on the oceans. The campaign uses stunning visuals, compelling stories, and actionable steps to engage the public and promote behavior change.

Another notable campaign is the BBC's "Blue Planet II," which highlights the beauty and fragility of marine ecosystems. The series not only educated viewers about the threats facing the oceans but also spurred policy changes and increased support for marine conservation initiatives.

CHAPTER THREE: EDUCATION AN CLIMATE LITERACY

3.1 Integrating Climate Change into School Curriculums

Education is a powerful tool for increasing climate awareness and literacy. Integrating climate change into school curricula ensures that students understand the science behind climate change, its impacts, and the importance of sustainability. Programs like the Next Generation Science Standards in the United States include climate change as a key component, promoting inquiry-based learning and critical thinking skills.

3.2 Higher Education and Research Initiatives

Higher education institutions play a critical role in advancing climate science and training the next generation of climate leaders. Universities offer specialized courses and degree programs in environmental science, climate policy, and sustainable development. Research initiatives, such as those conducted by the Earth Institute at Columbia University and the Grantham Institute at Imperial College London, contribute to our understanding of climate change and develop innovative solutions.

3.3 Community Education Programs

Community education programs, such as workshops, public lectures, and local sustainability initiatives, aim to raise awareness and empower individuals to take action. These programs often focus on practical steps that communities can take to reduce their carbon footprint and adapt to climate impacts. Examples include community gardens, energy efficiency workshops, and local climate action plans.

3.4 Online Courses and Resources

Online education platforms, such as Coursera, edX, and Khan Academy, offer courses on climate science, sustainability, and environmental policy. These courses make high-quality education accessible to a global audience and allow learners to engage with expert instructors and fellow students. Resources such as the Climate Literacy and Energy Awareness Network (CLEAN) provide educators with vetted teaching materials and activities.

CHAPTER FOUR: GRASSROOTS MOVEMENTS AND ACTIVISM

4.1 The Role of Non-Governmental Organizations (NGOs)

Non-governmental organizations (NGOs) play a vital role in climate advocacy and education. Organizations like Greenpeace, the Sierra Club, and the World Wildlife Fund (WWF) conduct research, run awareness campaigns, and mobilize public action. NGOs often serve as watchdogs, holding governments and corporations accountable for their environmental practices.

4.2 Youth-Led Movements (e.g., Fridays for Future)

Youth-led movements have brought fresh urgency to climate action. Inspired by Greta Thunberg, the Fridays for Future movement has organized global climate strikes, demanding action from policymakers. These movements leverage social media to organize events, share information, and engage with supporters worldwide.

4.3 Local and Global Climate Marches

Climate marches and demonstrations, such as the Global Climate Strike and Earth Day marches, bring millions of people together to demand climate action. These events raise public awareness, attract media attention, and put pressure on political leaders to implement climate policies.

4.4 Case Studies of Successful Activism

Examining successful activism campaigns provides insights into effective strategies and tactics. The Dakota Access Pipeline protests, led by Indigenous groups and environmental activists, highlighted the environmental and cultural risks of fossil fuel infrastructure. The divestment movement, which encourages institutions to divest from fossil fuels, has successfully shifted investments towards renewable energy and sustainable businesses.

CHAPTER FIVE: GOCERNMENT AND POLICY INITIATIVES

5.1 National Climate Policies

National governments play a crucial role in addressing climate change through policy and regulation. Policies such as the Clean Power Plan in the United States and the Renewable Energy Directive in the European Union aim to reduce GHG emissions and promote renewable energy. These policies set targets, provide incentives, and establish frameworks for transitioning to a low-carbon economy.

5.2 International Agreements (e.g., Paris Agreement)

International Increasing Public Awareness and Concern: Understanding and Addressing Climate Change

CHAPTER SIX: CORPORATE RESPONSIBILITY AND CLIMATE ACTION

6.1 The Role of Corporations in Climate Change

Corporations significantly contribute to greenhouse gas emissions, especially in industries like energy, transportation, agriculture, and manufacturing. As major stakeholders in the global economy, corporations have a critical role in mitigating climate change by reducing their carbon footprints, adopting sustainable practices, and innovating in clean technologies.

Corporations are increasingly recognizing the financial risks associated with climate change, such as disrupted supply chains, resource scarcity, and regulatory changes. This awareness is driving many companies to integrate climate considerations into their business strategies and operations.

6.2 Corporate Sustainability Initiatives

Many corporations are implementing sustainability initiatives to reduce their environmental impact. These initiatives include energy efficiency improvements, waste reduction programs, and sourcing renewable energy. Companies like Google and Apple have committed to running their operations on 100% renewable energy, setting an example for others to follow.

Sustainability reporting has also become more common, with companies disclosing their environmental performance and climate-related risks through frameworks like the Global Reporting Initiative (GRI) and the Task Force on Climate-related Financial Disclosures (TCFD). These reports provide transparency and accountability, helping stakeholders assess corporate progress on climate goals.

6.3 Innovations in Clean Technology

Innovation in clean technology is essential for transitioning to a low-carbon economy. Corporations are investing in research and development of technologies such as renewable energy, electric vehicles, energy storage, and carbon capture and storage. Companies like

Tesla and Siemens are at the forefront of developing and deploying these technologies.

The shift towards circular economy models, where resources are reused and recycled, is also gaining traction. Corporations are redesigning products and processes to minimize waste and extend the lifecycle of materials, contributing to resource efficiency and emissions reductions.

6.4 Case Studies of Corporate Leadership

Case studies of corporate leadership in climate action provide valuable insights into successful strategies and practices. For instance, Unilever has committed to achieving net-zero emissions across its value chain by 2039, with initiatives spanning sustainable sourcing, energy efficiency, and waste reduction.

Patagonia, an outdoor apparel company, integrates environmental sustainability into its core mission, promoting responsible consumption and advocating for environmental protection. The company donates a portion of its profits to environmental causes and encourages its customers to repair and reuse products instead of buying new ones.

Another example is IKEA, which aims to become climate positive by 2030. The company invests in renewable energy, improves energy efficiency in its stores and supply chain, and promotes sustainable living products.

CHAPTER SEVEN: INDIVIDUAL ACTIONS AND LIFESTYLE CHANGES

7.1 Reducing Carbon Footprints

Individuals can significantly contribute to climate change mitigation by reducing their carbon footprints. Simple actions like using energy-efficient appliances, reducing water consumption, and minimizing waste can collectively have a substantial impact. Opting for public transportation, cycling, or walking instead of driving can also reduce personal emissions.

7.2 Sustainable Consumption

Adopting sustainable consumption habits involves making conscious choices about what we buy and use. This includes selecting products with minimal packaging, choosing energy-efficient and durable goods, and supporting companies with strong environmental commitments. Reducing meat consumption and opting for plant-based diets can also lower greenhouse gas emissions, as livestock farming is a major contributor to methane emissions.

7.3 Renewable Energy and Home Improvements

Switching to renewable energy sources, such as installing solar panels or purchasing green energy from utilities, helps reduce reliance on fossil fuels. Home improvements, such as better insulation, energy-efficient windows, and smart thermostats, can enhance energy efficiency and reduce heating and cooling needs.

7.4 Community Involvement and Advocacy

Individuals can amplify their impact by engaging in community actions and advocacy. Joining local environmental groups, participating in climate marches, and supporting policies that promote sustainability are effective ways to contribute. Advocating for renewable energy projects, improved public transportation, and green spaces in local communities can drive broader environmental benefits.

7.5 Education and Raising Awareness

Education plays a crucial role in empowering individuals to take climate action. Learning about climate change and its impacts can inspire more sustainable choices. Sharing knowledge and raising awareness within social circles can create a ripple effect, encouraging others to adopt eco-friendly practices.

CHAPTER EIGHT: THE FUTURE OF CLIMATE ACT

8.1 Emerging Technologies

Emerging technologies hold promise for addressing climate change. Innovations in artificial intelligence, blockchain, and the Internet of Things (IoT) are being explored to optimize energy use, enhance renewable energy integration, and improve climate resilience. Advances in battery technology and grid management can facilitate the widespread adoption of renewable energy sources.

8.2 Global Collaboration and Policy Development

Global collaboration and policy development are essential for effective climate action. Strengthening international agreements, such as the Paris Agreement, and enhancing cooperation between countries can drive collective efforts to reduce emissions. Policies that promote renewable energy, carbon pricing, and sustainable development can provide the framework for meaningful progress.

8.3: Adapting to Climate Impacts

Adapting to climate impacts is crucial for building resilience. This includes developing infrastructure to withstand extreme weather events, implementing early warning systems, and supporting communities vulnerable to climate change. Integrating climate adaptation into urban planning, agriculture, and water management can enhance resilience and reduce risks.

8.4 Mobilizing Finance for Climate Action

Mobilizing finance for climate action is critical to supporting mitigation and adaptation efforts. Green finance mechanisms, such as green bonds and climate funds, can channel investments into sustainable projects. Public-private partnerships can leverage resources and expertise to drive large-scale climate solutions.

8.5 The Role of Individuals in Shaping the Future

Individuals play a vital role in shaping the future of climate action. Personal choices, community engagement, and advocacy can drive significant change. By staying informed, making sustainable choices, and advocating for stronger climate policies, individuals can contribute to a collective effort to address climate change and build a sustainable future.

CHAPTER NINE: INNOVATION AND TECHNOLOGY IN CLIMATE SOLUTIONS

9.1 Renewable Energy Advances

Renewable energy technologies are essential for reducing greenhouse gas emissions and transitioning to a sustainable energy future. Significant advances have been made in solar, wind, and hydroelectric power. Photovoltaic solar panels have become more efficient and affordable, leading to widespread adoption in residential, commercial, and industrial sectors. Innovations like bifacial solar panels, which capture sunlight on both sides, and perovskite solar cells, which offer high efficiency at low cost, are pushing the boundaries of solar technology.

Wind energy has also seen remarkable progress with the development of larger and more efficient turbines. Offshore wind farms, which can harness stronger and more consistent winds, are expanding rapidly, contributing to significant increases in wind power capacity.

Hydroelectric power, long a cornerstone of renewable energy, is being enhanced through the modernization of existing dams and the development of small-scale hydro projects. Innovations in pumped storage hydropower provide energy storage solutions by moving water between reservoirs at different elevations.

9.2 Energy Storage Solutions

Energy storage is critical for managing the intermittent nature of renewable energy sources. Advances in battery technology, particularly lithium-ion batteries, have enabled more efficient and affordable energy storage systems. These systems can store excess energy generated during periods of high renewable output and release it when demand is high or renewable generation is low.

Emerging technologies like solid-state batteries promise even greater energy density and safety, potentially revolutionizing energy storage. Additionally, flow batteries, which use liquid electrolytes, offer scalable storage solutions for large-scale applications.

Other energy storage methods, such as compressed air energy storage (CAES) and flywheel energy storage, are being explored for their potential to provide long-duration storage and grid stability.

9.3 Smart Grid Technologies

Smart grid technologies enhance the efficiency, reliability, and resilience of electricity grids. These technologies enable real-time monitoring and management of electricity flows, integrating renewable energy sources, and optimizing energy use.

Advanced metering infrastructure (AMI) allows utilities to collect and analyze data on energy consumption, enabling demand response programs that adjust energy use during peak periods. Grid management systems use artificial intelligence and machine learning to predict demand and optimize the distribution of electricity.

Microgrids, which are localized grids that can operate independently or in conjunction with the main grid, enhance energy resilience by providing backup power during outages and integrating distributed energy resources like solar panels and battery storage.

9.4 Carbon Capture and Storage (CCS)

Carbon capture and storage (CCS) technologies are crucial for reducing CO_2 emissions from fossil fuel-based power plants and industrial processes. CCS involves capturing CO_2 emissions at their source, transporting them to storage sites, and injecting them into underground geological formations for long-term sequestration.

Recent advancements in CCS include the development of more efficient capture technologies, such as solvent-based and membrane-based systems. Enhanced oil recovery (EOR) techniques, which use captured CO_2 to extract additional oil from depleted reservoirs, provide an economic incentive for CCS deployment.

Direct air capture (DAC) technologies, which remove CO_2 directly from the atmosphere, are also being developed. These technologies have the potential to offset emissions from sectors that are difficult to decarbonize, such as aviation and heavy industry.

9.5 Innovations in Agriculture and Land Use

Agriculture and land use contribute significantly to greenhouse gas emissions, but innovations in these sectors can help mitigate climate change. Precision agriculture technologies, such as GPS-guided equipment and drones, enable farmers to optimize inputs like water, fertilizer, and pesticides, reducing emissions and improving yields.

Regenerative agriculture practices, including cover cropping, no-till farming, and agroforestry, enhance soil health, sequester carbon, and increase resilience to climate impacts. Innovations in livestock management, such as feed additives that reduce methane emissions from ruminants, offer additional mitigation opportunities.

Forest conservation and reforestation initiatives are vital for carbon sequestration. Advances in remote sensing and satellite monitoring help track deforestation and land-use changes, enabling more effective conservation strategies.

9.6 Case Studies of Technological Innovations

Case studies of successful technological innovations illustrate the potential for transformative climate solutions. For example, Denmark's integration of wind power into its energy system demonstrates the feasibility of achieving high renewable energy penetration while maintaining grid stability. The country generates nearly half of its electricity from wind, supported by robust grid infrastructure and international energy cooperation.

The Tesla Gigafactory in Nevada exemplifies advancements in battery technology and large-scale manufacturing. The facility produces lithium-ion batteries for electric vehicles and energy storage systems, driving down costs and increasing production capacity.

In Kenya, the M-KOPA solar project provides affordable solar home systems to off-grid households, using mobile money platforms for payments. This innovation addresses energy access and sustainability, improving the quality of life for millions of people.

CHAPTER TEN: CLIMATE CHANGE AND PUBLIC HEALTH

10.1 Health Impacts of Climate Change

Climate change poses significant risks to public health, affecting various aspects of human well-being. Rising temperatures and changing weather patterns increase the frequency and severity of heatwaves, leading to heat-related illnesses and deaths. Vulnerable populations, such as the elderly, children, and individuals with pre-existing health conditions, are particularly at risk.

Changes in precipitation patterns and extreme weather events contribute to the spread of infectious diseases. For example, warmer temperatures and increased rainfall create favorable conditions for the proliferation of mosquitoes, spreading diseases like malaria, dengue fever, and the Zika virus. Flooding and hurricanes can contaminate water supplies, leading to outbreaks of waterborne diseases.

Air pollution, exacerbated by climate change, poses another significant health threat. Increased concentrations of ground-level ozone and particulate matter result in respiratory and cardiovascular illnesses, impacting millions of people worldwide.

10.2 Mental Health and Climate Anxiety

The psychological impacts of climate change are increasingly recognized as a public health concern. Climate anxiety, or eco-anxiety, refers to the chronic fear of environmental doom. Individuals, particularly young people, experience stress, depression, and anxiety related to the future impacts of climate change.

Natural disasters and extreme weather events can lead to post-traumatic stress disorder (PTSD), anxiety, and depression among affected communities. The loss of homes, livelihoods, and loved ones during such events has profound mental health consequences.

Addressing mental health impacts involves providing support services, promoting community resilience, and fostering a sense of agency and hope through climate action.

10.3 Strategies for Mitigating Health Risks

Mitigating health risks associated with climate change requires a multifaceted approach. Public health systems must be strengthened to respond to climate-related health threats, including by improving disease surveillance, enhancing emergency preparedness, and building healthcare infrastructure resilient to climate impacts.

Policies that reduce greenhouse gas emissions also offer significant health benefits. For example, transitioning to clean energy sources reduces air pollution, resulting in fewer respiratory and cardiovascular diseases. Promoting active transportation, such as walking and cycling, can improve physical fitness and reduce obesity rates.

Community-based adaptation strategies, such as early warning systems for heatwaves and flood risk reduction measures, enhance resilience to climate impacts. Public awareness campaigns and education programs can inform individuals about the health risks of climate change and encourage proactive measures to protect their health.

10.4 Case Studies of Public Health Responses

Examining public health responses to climate change provides valuable insights into effective strategies. In France, the implementation of a heatwave warning system, coupled with public education and emergency preparedness measures, significantly reduced heat-related deaths during subsequent heatwaves.

Bangladesh's community-based flood preparedness programs, which include early warning systems, evacuation plans, and health services, have improved resilience to flooding and reduced health impacts.

The Clean Air Act in the United States, which regulates air pollutant emissions, has achieved significant reductions in air pollution, leading to improved public health outcomes. The act demonstrates the effectiveness of regulatory measures in addressing environmental health risks.

CLIMATE ELEVEN: THE ECONOMICS OF CLIMATE ACTION

11.1 The Cost of Inaction

The economic costs of inaction on climate change are substantial and far-reaching. Failure to address climate change leads to increased costs from natural disasters, health impacts, and disruptions to agriculture and infrastructure. Extreme weather events, such as hurricanes, floods, and wildfires, cause billions of dollars in damage annually, burdening governments, businesses, and individuals.

Climate change also threatens economic stability by impacting critical sectors such as agriculture, fisheries, and tourism. Changes in temperature and precipitation patterns disrupt crop yields and fisheries, leading to food insecurity and economic losses. Coastal tourism is threatened by sea-level rise and extreme weather, affecting jobs and revenue.

11.2 Economic Benefits of Climate Action

Investing in climate action offers significant economic benefits. Transitioning to renewable energy creates jobs in manufacturing, installation, and maintenance. The renewable energy sector already employs millions of people worldwide and is expected to grow rapidly as more countries adopt clean energy policies.

Energy efficiency measures reduce energy costs for households and businesses, freeing up resources for other expenditures and investments. Retrofitting buildings, upgrading appliances, and improving industrial processes enhance competitiveness and reduce operating costs.

Climate action also fosters innovation and new business opportunities. Companies that develop and deploy clean technologies gain a competitive edge in the global market. Green finance, including sustainable investing and green bonds, attracts capital for climate-related projects, driving economic growth and environmental benefits.

11.3 Market-Based Solutions

Market-based solutions, such as carbon pricing and cap-and-trade systems, are effective tools for reducing greenhouse gas emissions. Carbon pricing assigns a cost to emitting CO_2, incentivizing businesses and individuals to reduce emissions and invest in cleaner technologies. Revenues generated from carbon pricing can be used to fund climate mitigation and adaptation projects, support vulnerable communities, and reduce other taxes.

Cap-and-trade systems set a limit on total emissions and allow companies to buy and sell emission allowances. This market mechanism ensures that emission reductions occur where they are most cost-effective. The European Union Emissions Trading System (EU ETS) and California's cap-and-trade program are examples of successful implementations of this approach.

www.ingramcontent.com/pod-product-compliance
Lightning Source LLC
Chambersburg PA
CBHW071959210526
45479CB00003B/998